"MEAT WITHOUT BONES"

is what the ancient Chinese called the versatile and healthful soy bean — and the soyfoods presented in this book are indeed excellent sources of protein, true replacements for meat.

The tasty tofu, the crisp tempeh, the pungent miso, provide a wide range of main dishes, side dishes, soups, sauces and desserts — with the health benefits of lowered plasma cholesterol, low-calorie/high-protein nourishment, and essential trace minerals.

For economy, culinary pleasure, and good health, you owe it to yourself to discover the facts about soyfoods.

ABOUT THE AUTHOR AND EDITORS

Richard Leviton is Editor/Publisher of *Soyfoods* Magazine, an international trade journal for producers of soyfoods. He is executive director of the Soyfoods Association of North America, the industry's trade group, which sponsors an annual international soyfoods technical conference and exhibition. He writes regularly on soyfoods for such publications as *New Age, East West Journal, In Business, Vegetarian Times, Whole Foods, Cooking for Survival Consciousness,* as well as numerous articles for *Soyfoods.*

Richard A. Passwater, Ph.D., is one of the most called-upon authorities for information relating to preventive health care. A noted biochemist, he is credited with popularizing the term "supernutrition" largely as a result of having written two bestsellers on the subject: *Supernutrition: Megavitamin Revolution* and *Supernutrition for Healthy Hearts.* His other books include *Easy No-Flab Diet, Cancer and Its Nutritional Therapies,* and the recently published *Selenium as Food & Medicine.* He has just completed a new book on *Hair Analysis* with Elmer N. Cranton, M.D.

Earl Mindell, R.Ph., Ph.D., combines the expertise and working experience of a pharmacist with extensive knowledge in most of the nutrition areas. His book *Earl Mindell's Vitamin Bible* is now a million-copy bestseller; and his more recent *Vitamin Bible for Your Kids* may very well duplicate the first Bible's publishing history. Dr. Mindell's popular *Quick and Easy Guide to Better Health (Vitamins Are Good for You!)* was just published by Keats Publishing.

TOFU, TEMPEH, MISO AND OTHER SOYFOODS

THE "FOOD OF THE FUTURE" — HOW TO ENJOY ITS SPECTACULAR HEALTH BENEFITS

by Richard Leviton

Keats Publishing, Inc. **New Canaan, Connecticut**

Tofu, Tempeh, Miso and Other Soyfoods is not intended as medical advice. Its intention is solely informational and educational. Please consult a medical or health professional should the need for one be warranted.

TOFU, TEMPEH, MISO AND OTHER SOYFOODS

ISBN: 0-87983-284-3

Printed in the United States of America

Good Health Guides are published by Keats Publishing, Inc.
27 Pine Street (Box 876)
New Canaan, CT 06840

CONTENTS

CONTENTS

THE NEIGHBORHOOD SOY DELI

Nutrition-conscious Americans today are looking for the fast tofu burger, for savory soysage on rye, with a side order of Tofuna salad, followed by a cone of soft-serve strawberry soymilk ice cream. While the neighborhood MacTofu's soy deli may not have opened yet in many communities, soyfoods — delicious, ready-to-eat, nutritious, inexpensive, convenient — are entering the American marketplace and stirring up a great deal of excitement.

The soy deli — an entirely American invention for retailing prepared soyfoods dishes made from fresh tofu, tempeh and miso — is in the forefront of the move to popularize and "Americanize" these once strictly East Asian foods. Today tofu is becoming as American a food as pizza and yogurt.

In any of the handful of American (and Canadian) soy delis now operating, one can happily order a spicy tofu burger, a frosty banana-soymilk shake, crisp tempeh garlic chips, hearty miso vegetable soup, and top it off with a smooth wedge of tofu-lemon cheesecake. All the traditional categories of restaurant fare — soups, entrées, salads, shakes, sandwiches, desserts — are made from soyfoods, and most often the concoctions can be enjoyed at pleasant wooden tables or packaged up to relish at home or at the office.

Soy deli dishes are for the entire family. The eager youngster tugs at his mother's sleeve for another carob soymilk smoothie, the young executives in their three-piece corduroy suits dive into tempeh burgers, the well-groomed matron delicately forks her kiwi tofu cheesecake, beaming grandmothers buy tofu and Finger Lickin' miso by the pound from the take-home case, and boisterous teenagers plow through another round of banana chip soymilk ice cream cones.

Soyfoods are turned into imaginative, tasteful, and irresistibly tempting dishes: spicy tempeh Reubens, steaming tofu pizzas, hearty tempeh pot pies, piquant tofu spinach-dill turnovers, rich tofu lasagna, chilled Tofelafels, creamy tofu mayonnaise potato salad, tangy deviled tofu salad, smooth tofu butternut squash pie, or sweet banana-coconut tofu cream pie — the list is nearly endless. But with each mouthwatering soyfood dish comes a balanced, adequate, and sustainable nutritional package.

Soyfoods, prepared as convenience foods, are nutritious, fast, satisfying, low-cost, easily duplicated in anyone's kitchen. The soy deli is simply a demonstration of the variety of dishes any kitchen gourmet can create at home with tofu, tempeh, and miso. The dairy industry often pictures a contented cow, its flanks dotted with illustrations of milk, yogurt, cheese, butter, and cream, like a map of the territory. Similarly, the soybean plant, dubbed the "cow of China" and the "meat without bones" in the Orient, will soon be portrayed in the popular imagination as the wonder bean replete with promises of marvelous food products.

One's first encounter with tofu might be a simple bowl of fresh warm curds topped with soy sauce, or tofu chunks simmering in a fragrant miso broth, or a creamy onion dip with crisp tempeh sticks. The 1980s will be known as the Decade of Soyfoods. As the marketing vanguard of soyfoods triumphant march into the American marketplace and onto the dinner plate, soy delis and convenience soyfoods are a certain way to win converts to the unmatchable blends of nutrition and excitement that are tofu, tempeh, and miso.

The soybeaning of the American kitchen is already under way. Now that the table is set and our appetites whetted, the centuries-old story of the soybean and its many foods needs telling.

WHAT ARE SOYFOODS?

The term "soyfoods" is a new word developed in the United States to describe the full range of edible human foods crafted from one versatile raw material, the soybean. Just as the dairy industry has created a universe of products based on the milk of the pasture-bound cow, today's whole array of soyfoods derives from one simple legume plant. Typically soyfoods are high in protein, low in calories, fats, and carbohydrates, free of cholesterol, easy to digest, inexpensive, versatile in the kitchen, delicious to eat, enjoyable to make, and exciting to talk about.

Tofu (pronounced **toe**-fu), probably the most popular soyfood today, is a white cake of pressed curds made from coagulated soymilk, much as cheese is made from milk. Known sometimes as "soybean cheese" or "bean cake," tofu is sold packed in water (in 8- and 16-ounce cakes), is somewhat bland in taste, and ranges in texture from custardy to firm. Three styles of tofu are distinguished: a soft, silken style which can be eaten with a spoon, a regular, more solid style, and a very firm and dense tofu. Freshly made tofu has an exquisitely sweet and subtle taste and aroma; regrettably much of this wonderful flavor disappears in a day after immersion in cold water.

Tofumaking is a simple process, as easily performed at home as breadmaking, and begins with soaking one pound of dry soybeans in water overnight. The next morning the beans have swelled to more than double their original size. They are measured with water into a blender and made into a fine puree, which is then heated on the stove for about twenty minutes. The hot liquid, which is strained off from the pulp, is called soymilk, a rich, frothy beverage suitable for immediate consumption. The pulp, called okara (pronounced oh-**kar**-rah) or "honorable hull" in Japan, is set aside.

The rich soymilk is coagulated with one of several possible curding agents (including natural or refined nigari, derived from sea water, calcium sulfate, calcium chloride — all naturally derived salts), and tofu curds and whey appear. The curds are pressed in tofu forming boxes until they cohere into a large uniform cake. The finished tofu may be consumed fresh or stored in water and will keep about two weeks refrigerated.

Tempeh (pronounced **tem**-pay) is a fermented soybean patty (generally an 8-ounce slab one-quarter inch thick) with a pleasantly nutty aroma and soft, chewy, almost meatlike texture. Tempeh is made from tender cooked soybeans which have been split and dehulled beforehand. The soft beans are inoculated with tempeh starter (a dry powder of the mold *Rhizopus oligosporous*), placed in perforated plastic bags, then incubated in a warm, moist environment for about twenty-four hours.

Finished tempeh has a lovely nap of white mycelium, as downy as the nap of new tennis balls, that holds the beans together. Tempeh can be stored refrigerated in a plastic bag and will keep about ten days. Often the mold continues to grow and black spots, then grey splotches, appear. This is normal and overripe tempeh such as this may be eaten and enjoyed without worry. Tempeh is also made from blends of soybeans and rice, soybeans and wheat, garbanzo beans alone, and from okara.

Miso (pronounced **mee**-soh) is a fermented seasoning, or condiment, somewhat salty in taste, with a consistency like peanut butter. Miso is made from soybeans, salt, water, often rice or barley (optional) and the miso starter culture called koji (made from polished rice inoculated with the mold *Aspergillus oryzae*).

Miso comes in numerous flavors, colors, levels of saltiness, variations in texture and aroma, as varied, in fact, as are wines and cheeses. Miso ranges in color from tan to rusty red to chocolate brown, and can take anywhere from one month to three years to ferment, depending on the ingredients. Rice, barley, and soybean miso are suitable for broth bases, while

Finger Lickin' miso (made with chopped vegetables to create a chunky texture), Sweet Simmered miso (full of peanuts, walnuts, sesame seeds, grated lemon, garlic or ginger, and simmered with sweeteners until thick) are used as toppings for grains and vegetables.

The miso-making process involves two independent preparations that come together in the end. First koji is prepared. Polished rice is steamed until tender, then inoculated with *Aspergillus oryzae* and set to ferment for a day in a koji bed. Then the kernels, partly bound together by the developing mold, are transferred to shallow trays for another twenty-four hour growth cycle. When they are bound together in large clumps, the misomaker breaks them apart, sifts them, and adds salt. The koji is ready.

Meanwhile soybeans are boiled (after an overnight soaking in water) until tender, then extruded through a meat grinder like spaghetti strands. The long moist strands are mixed with the koji, salt, and water. The miso paste is transferred to large vats, often 300 to 600 gallons in capacity, and left to ferment for many months. The miso fermentation breaks down the grains and soybeans and transforms them into a fragrant miso, while the salt acts as a natural preservative. Fresh rice miso, after a quarter year's fermentation, surprisingly resembles sweet smoked ham in taste.

Soy sauce is a generic term referring to three different but related products. Shoyu (pronounced **show**-yu) is a liquid, dark condiment made from wheat, soybeans, salt, water, and koji. Tamari (pronounced tah-**mah**-ree) is often mistaken (and mislabeled) for shoyu but contains no wheat, although otherwise it has the same ingredients. Fermentation for shoyu and tamari takes about one year. Soybeans and wheat are prepared similarly to the miso process, and are made into a mash left to ferment in a vat. The liquid that oozes out of the mash is traditionally called tamari, which in Japanese means "to accumulate." Chemical soy sauce, commonly found in supermarkets, is usually made in one day by using a chemical hydrolysis process.

While most of America's attention today is pinned on these

popular soyfoods, many others, though less well known, are worthy of mention. Yuba (pronounced **you**-bah) is the delicate skin that forms on gently heated soymilk and is scooped up to stretch-dry. Natto (pronounced **nah**-toh), or fermented whole soybeans (made with *Bacillus subtilus*), has a sticky, slippery surface with fine filaments that hold the darkened beans together. Natto is used as a rice condiment or in soups, and is often served powdered.

Kinako (pronounced **keen**-ah-koh) is a delicious brown powder made from dry roasted soybeans and is used in confections. Soy viili (pronounced **veeh**-lee), new on the soyfoods scene, is a delightful, stretchy, yogurtlike product made from soymilk and the Finnish bacterial culture *villi*. Soy nuts — dry roasted or oil roasted, then seasoned, soybeans — make a delicious snack or peanut substitute in baking. Soy sprouts — long, nutritious green tendrils made from soaked soybeans — are tasty in wok-fried vegetable dishes. Green soybeans, steamed in the pods, taste exactly like buttered lima beans and can become addictive.

Still other basic soyfoods can be listed. Fermented tofu, called *toufu-ru* (or sometimes sufu) is tofu fermented with *Actinomucor elegans* which envelops the tofu cake with a fragrant white mycelium. The transformed tofu is immersed in a brine to age for up to six months, when it has a soft cheeselike consistency. Deep-fried tofu pouches (in which small thin chunks of tofu puff up like pita bread), deep-fried tofu cutlets (thick slabs of tofu deep-fried until crisp), deep-fried tofu burgers (tofu mashed with chopped vegetables, then deep-fried), and dried-frozen tofu (fresh tofu is frozen, thawed, oven-dried, to form powder-dry tofu cakes) — are but a few more of the soybean's marvelous progeny.

Not to be overlooked are soy flour (used widely as a protein booster in baking), soy oil (one of America's predominant salad oils), soy flakes (used in granolas), high-protein isolated soybean powders (called isolates and concentrates), and a variety of soybean meat analogs (textured vegetable proteins).

This brief list of soyfoods only touches on the primary products and doesn't elaborate on the exciting, possibly inexhaustible, realm of secondary foods made from them.

SOYBEANS FROM CHINA

Today's soyfoods, while they are exciting and important news for Westerners, are actually not recent innovations but have formed the protein backbone of the East Asian diet for several thousand years. Soybeans, although originally domesticated in China, have been big agricultural business for American farmers since 1920 and today are one of America's most lucrative farm crops. As soyfoods grow in popularity in the United States, the soybean's millennial East Asian message, that soybeans belong on the dinner plate and not in the feed trough, is being heard.

The U.S. soybean crop is vast yet many outside the Midwest will be astonished to learn that this $14 billion crop occupies 71.6 million acres, with an annual yield of 2.26 billion bushels in 1979, 1.81 billion in 1980, and 2.07 billion in 1981. A bushel of soybeans is 60 pounds, so that's 135 billion pounds of soybeans for 1979.

Soybeans are grown on a large scale in about twenty-four states, from Arkansas to Minnesota along the Mississippi River valley, and along the Eastern seaboard from Florida to New Jersey and New York (but not New England). Illinois and Iowa, the nation's leading producers, have combined soybean sales of $4.7 billion annually.

The United States is by far today's major world supplier of soybeans, outproducing its nearest competitor, Brazil, by almost three times. Argentina, China, Canada, and Mexico are other soybean-growing nations. Total world soybean harvests for 1980 were 81 million metric tons (or 180 billion pounds), of which some 45 percent worldwide went for human foods, mainly in East Asia.

In its three thousand years of known history, the modest soybean has worked its slow evolutionary path from mainland China to the American Midwest. Soybeans (of which there are some 8500 different genetic strains) belong to the genus *Glycine*, which in turn has two subgenera: *Glycine max*, the cultivated soybean, and *Glycine soja*, its wild relative. Agronomists believe that domestication of the wild soybean had been completed by about 1100 B.C. in east central China, which has since been called the soybean's primary gene center. The soybean migrated with merchants, missionaries, soldiers, and travelers to Japan, Korea, and southeast Asia beginning in 100 A.D., and then on to Europe (circa 1500).

Enough evidence exists, in fact, to implicate our own American folk hero, Benjamin Franklin, with the soybean's final move from the *Jardin des Plantes* in Paris to Philadelphia in the late 1780s. Franklin was a member of the Philadelphia Society for Promoting Agriculture and ambassador to France, where soybeans had been cultivated in small plots since 1739. According to Dr. Theodore Hymowitz, Franklin arranged for samples to be shipped to him and since then, the soybean has worked its way into agricultural significance and public prominence.

By as early as 1890, soybeans were planted commercially in Illinois and North Carolina, and by 1920 groups of farmers were holding annual "soybean days." During these festive weekends, they shared information and excitement about what is now called the Cinderella crop. USDA scientists made early research trips to China and Japan to collect soybean samples and to study indigenous soyfoods. Through the inspired pioneering work of numerous farmers and researchers, soybeans grabbed hold of the American plow and set it furrowing a new field.

The use of soybeans as human foods, however, is a relatively new trend in America. Soybeans traditionally have been used for livestock feed (from the protein-rich meal) and vegetable oil. Exporting soybeans is a major business, with about 57 percent of the annual harvest being sold abroad to help settle America's balance of payments. About one billion bushels go into "the crush," which refers to the basic processing of soybeans into meal and oil.

In 1981 tofu, tempeh, miso, and other soyfoods accounted for only 2.1 percent of this massive crop. Yet this amount, although it seems meager, is historically a high-point in the West for direct soybean consumption and it shows that the soybean, our Chinese import, is finding its true use as a food expander.

SOYFOODS HAVE MARCHED WEST

While the origins and early days of tofu, tempeh, and miso are not as easy to trace as their well-documented latter-day westernization, occurring now, their cradles have been China, Japan, and Indonesia. So highly valued were the tofu and miso in Japan that they became part of the national haute cuisine, the emperor's choice foods.

Tofu, many believe, was the discovery (sometime around 164 B.C.) of a Chinese alchemist, Taoist, and prince, Lord Liu An of Huia-nan. A dabbler in foods and transformations, it's speculated that one day he (or his cook) dropped natural salt into a pot of simmering soybean puree and was astonished to see curds form as a new food was revealed. Other scholars suggest tofu might have been imported by neighboring tribes, such as Indians to the south or Mongols to the north.

By 700 A.D. tofu had reached Japan, borne largely on the culinary backs of mendicant Buddhist monks who relied on tofu as their vegetarian protein staple. These monks became instrumental in tofu's dissemination throughout Japan; they instructed laymen in its preparation and served tofu dishes at their temple restaurants. By 1336 tofu was spreading outward from the major cities and into the countryside, no longer only a privileged food for princes and monks. Tofu had become Japanese, changing in nature from the Chinese, firmer style.

In Japan today the tofu industry is as pervasive as breadbaking is in America. There are some 28,000 tofu shops, mostly small, producing millions of pounds of fresh tofu. Twelve hundred years after tofu reached Japan, it arrived in America in San Francisco, where (around 1915) the first known Asian-American shop opened. And today, nearly seventy years later, America has over 170 active tofu shops.

Miso derives from an earlier fermented food called *chiang* (pronounced **channg**) which some believe to be man's earliest condiment. Chiang, developed sometime before 700 B.C., was devised as a way to preserve high-protein foods such as fish and meat during the winter without spoilage. Chiang had an applesauce or chunky porridge-like consistency.

Later in time, soybeans were substituted for the animal ingredients and miso was born. Numerous flavors, colors, and consistencies emerged and misos took the name of the region in which they were produced, as in Hatcho miso. Buddhist monks brought chiang to Japan sometime between 540 and 663 A.D., and by 950 misomaking had spread into the Japanese countryside. Today over 2500 miso craftsmen work daily at miso production, mostly in the traditional small-scale manner.

Tempeh came from West Java, Indonesia, and while scholars believe it originated as much as two thousand years ago, the only certain facts they have places its appearance at about 1750 A.D. Today tempeh is made at 41,000 Indonesian shops, although it didn't migrate off the primary island until the twentieth century. In the United States today there are already forty tempeh companies.

Since the day Benjamin Franklin requested samples of the curious soybean from the Parisian agronomist, the soybean's food uses have aroused a steady stream of interest and experimentation in the West. Surges of interest in developing the soybean for human consumption have crested and fallen in the last two hundred years, yet each surge has come on stronger and lasted longer. By the 1930s the use of soybeans as animal feed was catapulted into economic significance when certain crucial nutritional barriers were overcome through better processing.

Over the decades, soyfood pioneers in the West have attempted to develop soyfoods, taking out patents on soymilk, soybean burgers, soy flour blends, and publishing numerous technical and popular papers. Madison College, a Seventh-day Adventist college in Tennessee, between 1910 and 1950 made remarkable progress in introducing a line of canned (and fresh) soyfoods, including green soybeans, seasoned tofu, soymilk, soy coffee, and soy meat analogs. Interest in soyfoods crested during the World War Two years when meat supplies dwindled. Postwar prosperity diminished this once-great interest in soyfoods as Americans turned to meat as affluence increased.

But by the early 1970s, the cultural climate had finally ripened for a long-term acceptance of these wonderful soyfoods. Another religious group in Tennessee, known as The Farm, pioneered the contemporary popularization of soyfoods by farming hundreds of acres of soybeans, then turning them into nutritious foods for their twelve hundred residents. Then in 1975 William Shurtleff published a milestone work, *The Book of Tofu*, which clearly and with enthusiasm gave the full story of tofu, along with five hundred recipes and the exhortation to "send tofu in the four directions."

Thus was the American tofu industry born. Today, a mere seven years later, tofu is found in nearly all supermarkets and groceries, and is well on its way to becoming a household staple.

THE NUTRITIONAL SIDE OF SOYFOODS

Millions of Americans today are actively concerned with the relations between diet and disease, between nutrition and well-being. Trends such as a widespread anxiety about obesity, the upsurge of consumer interest in light foods, calorie counting, and concern about cholesterol levels are well documented. Geriatric dietary needs, medically restricted diets, weight-reducing plans are prescribed daily.

Americans are voluntarily curtailing red meat consumption in favor of lighter meats and even occasional forays into meatless diets. Physical fitness, now linked with good nutrition, has assumed paramount importance. And the growing interest in gourmet cooking, ethnic foods, the renewed respect and clamor for "real" foods with quality and craftsmanship, are also coming on strongly.

Soyfoods have arrived in the American marketplace at a propitious moment. That they are uniformly high in protein but low in calories, carbohydrates, and fats, entirely devoid of cholesterol, high in vitamins, easy to digest, tasty, and wonderfully versatile in the kitchen, positions them as irresistible new food staples for the evolving American diet.

Consumer columnists frequently report that soybeans are "incomplete" or "unbalanced" proteins that cannot be solely relied on for one's daily protein needs. This is a regrettable misconception and out of alignment with what current research in the field of soy protein and human nutrition is revealing.

The contention is usually made that the soybean lacks adequate amounts of the amino acids methionine and cystine, and is therefore called "limiting" in these components. Protein quality is a measure of how well a protein source can meet the specific human amino acid requirements, based on an ideal protein such as eggs or milk, whose amino acid configuration is believed to correspond most closely to the body's demands.

What is not often cited in the popular accounts, but comes under increasing criticism in nutritional circles, is that the validity of the cornerstone measurement of protein quality itself has come under considerable question. This measurement is called PER (Protein Efficiency Ratio, or how much weight is gained compared to how much protein is consumed) and is based on rat feeding trials. The problem is that rats have much higher biological demands for methionine and cystine (often twice as much) than humans, yet the human requirements have been based on these extrapolated rat data. Humans are told, as a result, that certain of their amino acid requirements are a lot higher than they probably are and that, based on this, soybeans do not contain enough methionine and cystine for their needs.

A second critical current in protein discussions today centers around the high protein intake standards known as RDAs (Recommended Dietary Allowances). RDAs for protein (as set by the Food and Nutrition Board) state that a 154-pound man requires 56 grams, and a 129-pound woman, 46 grams of protein every day. Yet it's becoming widely believed that Americans consume about two times as much protein as their bodies really need and that the RDAs are too high. Excess protein is transformed into fat and bodily plumpness reigns. While individual protein needs vary, roughly 50 percent of the RDAs for adults have been suggested as suitable for normal activity.

The conclusions thus far show that soy protein quality is higher than was commonly thought and that humans require less protein than they normally consume. The third important point is that in a sensible varied diet ample protein is obtained through protein complementarity. For example, 2¼ ounces of tofu combined with one cup of brown rice produces a meal with 32 percent more protein than if each food were consumed by itself.

The nutritional climate surrounding soy protein is changing and soy is emerging as a far better, more balanced, and sustainable protein source than anyone had believed before. And not only is soy protein acceptable as a primary protein source, it also has specific positive benefits when consumed regularly.

Cholesterol dominates everyone's thoughts today. It is commonly known that soybeans (and all soyfoods made without animal ingredients) do not contain cholesterol. It's not so widely known that soybeans actually lower existing plasma cholesterol levels in humans, as experiments in Canada, Italy, West Germany, France, and the U.S. have shown.

The result of this ability of soy protein to lower cholesterol levels in the blood is called hypocholesterolemia. Among forty hospital patients who were fed a restricted soybean-based diet for three weeks, a twenty-percent decline in plasma cholesterol levels was noted. In another study, a diet mixture containing soy protein, apple pectin, and wheat bran was fed to humans; serum cholesterol levels dropped from 190 to 160 mg/100 ml after only fourteen days. For cholesterol watchers, this is surely welcome news.

Nutritionists have studied the soybean in the last few decades with an eye toward maximizing its nutritional qualities. In the process, they characterized various "antinutritional" factors, then developed methods to remove them. Trypsin inhibitors, for instance, are soybean enzymes that interfere with protein digestion in rats, cows, and pigs, causing growth inhibition. While trypsin inhibitors have not been shown (through actual human experimentation) to have an impact on humans, the standard cooking of soybeans that occurs in the production of all soyfoods eliminates up to 90 percent of the inhibitors. Heat treatment of soybeans also greatly improves the PER and general protein digestibility.

Flatulence can occur after consuming uncooked (or undercooked) soybeans. The human intestine lacks the necessary digestive enzyme, called alphagalactisidase, which metabolizes several complex sugars (or oligosaccharides) in soybeans. Again, cooking the soybeans eliminates nearly all of these sugars. Tofu and tempeh are essentially devoid of any flatus-producing activity.

Another area nutritionists are looking into with soybeans is in mineral absorption (specifically zinc and iron). Definitive and uncontested studies are yet to be conducted but the general sentiment is that soy protein does lower somewhat, or inhibit, zinc absorption. Phytate (or phytic acid) is the form in which 70 percent of the phosphorus in soybeans appears, and the phytate forms an insoluble bond with dietary zinc and calcium, lowering their absorption rate. The dietary significance of these lowered absorption rates is yet to be established.

One of Japan's prestigious daily newspapers reported recently that Japanese who consume miso soup daily develop up to 33 percent less stomach cancer than nonconsumers, and about 8 to 18 percent less than those who drink miso soup only sometimes. Miso soup was also shown to have marked effects in lowering susceptibility to stomach ulcers, heart and liver disease, and is believed to help remove radiation from the body as well.

Tofu has clinical applications in special, medically-restricted diets such as diabetes, hypoglycemia, lactose intolerance, heart disease, and atherosclerosis. No curative effects are claimed, but the advantage lies in tofu's nutritional profile and how it matches the special needs of these medical conditions.

With this general nutritional background sketched for soyfoods, we can focus our attention specifically on the individual merits of tofu, tempeh, and miso.

Tofu. Tofu is valued chiefly for its high protein, which ranges from a low of 4 percent on a wet basis (in the custardy, silken style) to a high of 15 percent (in the extra firm or dense style). Tofu has one of the lowest ratios of calories to protein found in any known plant food; 1 gram of usable protein is accompanied by only twelve calories. Tofu is 95 percent digestible, an excellent weaning or geriatric food. Its Net Protein Utilization (or NPU, an index of how much of a protein can actually be assimilated by the body, as compared to eggs, the standard, at 95) is 65, exactly the same as chicken, and only slightly lower than beef at 67. A 4-ounce serving of tofu supplies up to 25 percent of the adult RDA for protein, 20 percent of the iron, and 20 percent of the phosphorus.

Tempeh. Fresh tempeh has even more protein than tofu, with 19.5 percent, or about as much as beef and chicken (in quantity) and 50 percent more than hamburger. Tempeh has the highest quality protein (measured in PER) of any soyfood, at 2.43, compared to milk at 2.81; a wheat-soy tempeh has a PER of 2.79.

Tempeh, because it is fermented, is easy to digest, rated at 86 percent, and has a NPU of 56 (for plain soybean) and 76 (for wheat-soy).

One strong attraction tempeh has for vegetarians is its high level of the essential vitamin B12, commonly thought unavailable in the vegetable world. A 4-ounce serving of tempeh may contain as much as 160 percent the adult RDA for B12. Tempeh contains only 157 calories per 100 grams, making it excellent for dieters.

Miso. As with tofu, the protein content of miso varies with the style and recipe, ranging from as high as 20 percent to an average of 13. Miso contains small amounts of vitamin B12 and is rich in enzymes and lactic acid bacteria (as in yogurt). As a concentrated protein source, miso's NPU is 72, so that 100 grams supplies more usable protein than an equal weight of hamburger. Miso, which is usually 12 percent salt, reduces salt intake while adding flavor to the diet. Like tofu, miso has one of the lowest ratios of calories to protein, with as little as 11 calories per gram of protein.

Soyfoods clearly have the nutritional profile to become the daily protein backbone of the American meatless or conventional diet. Soyfoods such as tofu, tempeh, and miso, consumed in normal quantities along with grains, vegetables, and other foods, can supply excellent and sustainable nutrition.

SOYFOODS IN YOUR KITCHEN

One of the most impressive features of soyfoods is their unmatched versatility in the kitchen. Tofu, tempeh, and miso can fit into any dietary or ethnic foods niche and yield countless culinary surprises. They cooperate handsomely with conventional protein foods (such as dairy, eggs, meat, fish) as extenders or in combination dishes. Cooking with soyfoods appeals to the inexperienced, perhaps disinterested, fast-food-minded kitchen visitor as well as to the full-blown, aproned, capped, gourmet and wok habitué, and to all manner of epicures in between.

Tofu is probably the most versatile of our highly adaptable group of soyfoods. Tofu, of course, is best enjoyed fresh. It is excellent as a blended dairy product replacer in creamy vegetable dips, (onion or dill), in puddings, custards, and pies (strawberry, banana), mayonnaise, whipped cream, dressings, cake icings, even ice cream confections.

Tofu can be chunked for soups, casseroles, wok-sautéed vegetable dishes, deep-fried to form pouches (to stuff with grains and diced vegetables), mashed with vegetables and deep-fried to make burgers, marinated and then deep-fried to make cutlets. Tofu can be used in baking as an ingredient, scrambled with onions (like eggs), mashed to create eggless egg or tofunafish salads. Tofu can also be griddled with soy sauce, or frozen, then thawed, marinated, and dropped in soups, or deep-fried.

Simply to recite a few of the hundreds of possible dishes from tofu can be a mouthwatering experience. Are you interested in: carrot-raisin-walnut salad with tofu; tofu, peanut butter, and banana spread; deep-fried tofu with pineapple sweet/sour sauce; tofu-stuffed mushrooms; tofu shrimp dip; tofu spinach quiche; tofu dill rolls; tofu pancakes; barbecued tofu; tofu lasagna; or rich tofu chocolate mousse?

Recipe

Tofu Guacamole
Makes 1½ cups

½ pound regular tofu
1 ripe avocado, peeled
½ teaspoon salt
1 clove fresh garlic, crushed and minced, or ½ teaspoon
   garlic powder
2 tablespoons lemon juice
3 tablespoons mayonnaise
   Dash Tabasco sauce
   Paprika

Drain tofu and mash together with peeled avocado. Add remaining ingredients except paprika and mix well. Sprinkle paprika on top. Serve with taco chips and fresh vegetable slices.

(Recipe reprinted with permission from *SoyDairy Delights of Tofu*, Greenfield, MA: Soy to the World Publishing Co., 1981.)

As a further indication of tofu's many surprises, 1981 saw the world's first Tofu Cheesecake Bakeoff in Colorado, sponsored by the Soyfoods Association of North America. Ten luscious tofu cream pies topped variously with kiwi, strawberries, pineapple, or walnuts, were featured. Seventeen delighted judges forked their way through the toothsome array of tofu cuisine.

Tempeh, usually deep-fried, becomes crisp and golden brown, with a satisfying, hearty taste often compared to that of southern fried chicken, veal cutlets, or fish sticks — a real meat substitute. Tempeh serves well as a main course replacement for poultry and fish because of its texture, chewiness, and rich taste. It is often baked or broiled, deep-fried, shallow-fried, or sautéed, or marinated and steamed. Fresh tempeh has a succulent aroma much like mushrooms, while frozen tempeh keeps for months without any change in texture.

There are hundreds of delectable dishes using tempeh. Can you resist: tempeh shish kebab; tempeh, lettuce, and tomato sandwich; coriander and garlic crisp tempeh; savory tempeh in tomato-herb sauce; tempeh burgers; tempeh mock chicken salad; tempeh cacciatore; applesauce-topped tempeh chops; barbecued tempeh; deviled tempeh dip; tempeh potato pancakes; tempeh avocado sandwich; tempeh cabbage rolls; or how about crisp, deep-fried tempeh sticks with a creamy tofu dill and onion dip?

Recipe

"Nearly Chicken" Tempeh Salad
Makes 2 cups

6 ounces tempeh
1/3 cup celery, chopped
¼ cup parsley, chopped
1 green onion, chopped
1/3 cup mayonnaise
2 teaspoons prepared mustard
1 teaspoon shoyu, or ¼ teaspoon salt
¼ teaspoon turmeric
¼ teaspoon garlic powder
Pinch paprika

Cut tempeh into 1/3-inch cubes and steam in a vegetable steamer for 20 minutes; cool. Put vegetables in large bowl. Mix mayonnaise with next 5 ingredients. Add tempeh to vegetables, stir in dressing, and chill.

(Recipe printed with permission from *The Soy of Cooking,* by Reggi Norton and Martha Wagner. Eugene, OR: White Crane Publications, 1981.)

Miso — savory, rich-bodied, pleasantly varied, colorful — can be used like bouillon or meat stocks in soups (as in miso vegetable soup), as Worcestershire in sauces, dips, and dressings (such as tahini-miso vegetable gravy), as a chutney or relish, as a vegetable or grain topping, and as vinegar in pickle-making. Miso keeps for many months without refrigeration and spoilage; however, white mold will appear on the surface after a time. This mold can be scraped off and the remainder of the miso enjoyed without difficulty.

A miscellany of appetizing recipes would include: mushroom miso sauté; miso creamcheese dip; floating cloud miso dressing; miso-sesame-avocado spread; deep-fried tofu and wakame seaweed miso soup; scrambled eggs with miso, onions, and tofu; grilled corn-on-the-cob with miso; tofu-miso pâté; apple wedges with orange-miso topping; sliced bananas with peanut-butter-miso topping; guacamole with miso.

Recipe

Tomato Hors d'Oeuvres with Sesame Miso
Serves 8 to 12

¼ cup sesame butter or tahini
2 teaspoons miso
2 tablespoons mayonnaise
1 clove garlic, crushed
½ teaspoon honey
1 teaspoon sake or white wine
1 to 2 teaspoons water (optional)
3 tomatoes, cut into ½-inch-thick slices

Combine the first seven ingredients, mixing well. Arrange tomato slices on a serving platter and top each slice with a dollop of the sesame miso.

(Reprinted with permission from *The Book of Miso*, New York: Ballantine Books, 1981.)

Cooking with tofu, tempeh, and miso in your home should no longer be a mystery or a source of perplexity, and not at all difficult. Some thirty-five soyfood cookbooks are now available in the United States and it seems as if a new volume of recipes appears monthly.

Increasingly, new recipes incorporate tofu, tempeh, and miso into mainstream, conventional American dishes. Tofu cookbooks portray tofu with cheese and egg dishes, as fish or chicken extenders. Backup ingredients are no longer exotic but employ commonly used kitchen spices and herbs. The new recipes Americanize soyfoods but this, ironically, places them once again in the multiethnic cooking pot of American cuisine, including Mexican, Chinese, Italian, Greek, and French styles.

THE FUTURE

Tofu, tempeh, and miso have moved out of the Orient to find a place at the American table. Far from being a fad of the moment, soyfoods are America's new sensible food staple — versatile, inexpensive, nutritious, and convenient protein sources. The rapid and steady appearance of American-style soyfood cookbooks and the widespread availability of products in stores testifies to their growing penetration of the marketplace.

The supply of tofu, tempeh, and miso in the American food system grows at an equally fast clip. The popularity of appetizing convenience products — instant miso soup, frozen tofu lasagna, marinated tempeh burgers — and the proliferation of fast-serve soy delis guarantee that the steady supply of delicious soyfoods will not lag.

Many analysts speculate that tofu could become the "yogurt of the 1980s," referring to the dairy industry's wonder child whose sales grew enormously since 1968, when fruit flavors appeared. Others ask if soyfoods could become another dairylike industry someday.

Whether you make your own subtly sweet tofu, savory Finger Lickin' miso, or fragrant okara tempeh at home, or purchase soyfoods at the store, your delight in cooking and enjoying soyfoods daily will be incomparable. As a soyfood enthusiast, you will find new ways to use tofu, tempeh, and miso. Soyfoods, once dubbed "the foods of the future," are here now as the foods of today.

BIBLIOGRAPHY

Caldwell, B.E., ed. *Soybeans: Improvement, Production and Use*, Madison, WI: American Society of Agronomy, 1973.

Carroll, K.K. "Soya Protein and Atherosclerosis" *Journal of the American Oil Chemists Society* (March 1981), pp. 416–418.

Erdman, J. "Effects of Soya Protein on Mineral Availability," *Journal of the American Oil Chemists Society* (March 1981), pp. 489–492.

Fox, Madeline, "Tofu and Special Diets," *Soyfoods* 3 (1981), p. 9.

Immerman, Alan, "Protein: How Much Is Enough?" *Vegetarian Times* (April 1980), pp. 60–61.

Johnson, Gil, "Soyboom," *Natural Foods Merchandiser* (September 1981), pp. 9–33.

Kennedy, Melissa, ed. *Soya Bluebook 1981*. St. Louis, MO: American Soybean Association, pp. 135–172.

Leviton, Richard, "On the Road for the Fast Food Burger: A Soy Deli Odyssey," *Vegetarian Times* 56 (April 1982), pp. 62–65.

———, "Ted Hymowitz: Soybean Sleuth" *Soyfoods* 5 (1981), pp. 61–63.

Rackis, J.J. "Significance of Soya Trypsin Inhibitors in Nutrition," *Journal of the American Oil Chemists Society* (March 1981), pp. 495–500.

——— "Flatulence Caused by Soya and Its Control Through Processing," *Journal of the American Oil Chemists Society* (March 1981), pp. 503––509.

Shurtleff, William, *The Book of Tofu*. New York: Ballantine Books, 1979.

——— *The Book of Tempeh*. New York: Harper & Row, 1979.

——— *The Book of Miso* New York: Ballantine Books, 1981.

——— "Miso Soup — Safeguard Against Cancer," *East West Journal* (January 1982), pp. 42–43.

Smith, Alan, and Circle, Sydney, eds. *Soybeans: Chemistry and Technology, Vol. 1, Proteins*, Westport, CT: AVI Publishing, 1978.

Wilcke, Harold, Hopkins, Daniel, and Waggle, Doyle, eds. *Soy Protein and Human Nutrition*, New York: Academic Press, 1979.

AVAILABLE SOYFOODS COOKBOOKS

Aihara, Herman and Cornelia, *Soybean Diet*. Oroville, CA: George Ohshawa Macrobiotic Foundation, 1974.

Andersen, Juel, *The Tofu Primer*. Brattleboro, VT: Stephen Greene Press, 1978.

Juel Andersen's Tofu Kitchen. New York: Bantam Books, 1981.

Andersen, Juel, and Cathy Bauer, *The Tofu Cookbook*. Emmaus, PA: Rodale Press, 1979.

Clarke, Christina, *Cook With Tofu*. New York: Avon, 1981.

Cloud, Maxine, *The Soysage Cookbook*. Hyde Park, VT: Cloud Mountain Publishing, 1979.

Chen, Philip, *Soybeans for Health and a Longer Life*. New Canaan, CT: Keats Publishing, 1973.

Cooking with Tofu. Charlotte, VT: Garden Way Publishing, 1981.

Cherniske, Stephen, *Tofu — Everybody's Guide*. E. Woodstock, CT: Mother's Inn Center, 1980.

Farr, Barbara, *Super Soy*. New Canaan, CT: Keats Publishing, 1976.

Ford, Richard, *The Soy Foodery Cookbook*. Santa Barbara, CA: 1981.

Fox, Madeline, Kathleen O'Connor and Judith Timmins, *Delights of Tofu*. Greenfield, MA: Soy to the World Publishing, 1981.

Hagler, Louise, *The Farm Vegetarian Cookbook*. Summertown, TN: The Book Publishing Co., 1978.

Tofu Cookery. Summertown, TN: The Book Publishing Co., 1982.

Heartsong, Toni and Bob, *The Heartsong Tofu Cookbook*. Miami, FL: Banyan Books, 1977.

Hobson, Phyllis, *The Soybean Book*. Charlotte, VT: Garden Way Publishing, 1978.

Immegart, Mavis and Dansly, Patty. *The Incredible Tofu Cookbook — California Style*. Yorba Linda, CA: 1981.

Jones, Dorothea Van Gundy, *The Soybean Cookbook*. New York: Arco Publishing, 1963.

Kushi, Aveline, *How to Cook with Miso*. Tokyo: Japan Publications, 1978.

Landgrebe, Gary, *Tofu Goes West*. Palo Alto, CA: Fresh Press, 1978.

Tofu at Center Stage. Palo Alto, CA: Fresh Press, 1978.

Lemly, Virg and Jo, *Soybean Cookery*. Cave Junction, OR: Wilderness House Publications, 1975.

McGruter, Patricia, *The Great American Tofu Cookbook*. Brookline, MA: Autumn Press, 1979.

Nofziger, Margaret, *Tempeh Cookery*. Summertown, TN: The Book Publishing Co., 1982.

Norton, Reggi and Wagner, Martha, *The Soy of Cooking*. Eugene, OR: White Crane Publications, 1981.

Olszewski, Nancy, *Tofu Madness*. Vashon, WA: Island Spring Inc., 1978.

Omura, Yoshiake, *The Tofu-Miso High Efficiency Diet*. New York: Arco Publishing, 1981.

Sheppard, Sally, *Tofu Cookbook*. Salt Lake City, UT: Jack's Beanstalk, Inc., 1981.

Uhlinger, Susan, *Soybean Cooking*. Brattleboro, VT: Stephen Greene Press, 1978.

Wolf, Ray, Ed, *Home Soyfood Systems*. Emmaus, PA: Rodale, Press, 1981.

KEATS GOOD HEALTH GUIDES

25 Titles in Print or Preparation...

Editors: Richard A. Passwater, Ph.D. and
Earl Mindell, R.Ph., Ph.D.